MW01502647

Searching for Cloud Nine

A guide to Cloud Computing for Businesses

By Mark Putiyon

<u>Dedication</u>

I would like to dedicate this book to my family.

To my wife Kim who has shown me, <u>not</u> told me, that with a disciplined mind you can achieve anything. To Mom and Dad, you have always praised me for my accomplishments and consoled me in defeats. Little do you know how much your positive affirmation has molded me into the man I am today. The humor in this book was inspired by my four boys, Matthew, Stephen, Joseph and Jonathan. They keep telling me "Dad you are not funny!" #dadjokesRlame. To "My Jennie" in your best Forrest voice, while not blood relation, I consider you family. To Ricky, Jeff and Darlene my bros and sis thanks for keeping me grounded and man when we get together..whoa !!! To my Uncle John who showed me how to joke around.

But mostly to JC – because He is The Man!

Disclaimer

All ideas, views and thoughts expressed in this book are the author's own. References have been provided wherever possible. *Searching for Cloud Nine* is not affiliated, authorized or endorsed with any of the brands and names mentioned in here unless specified otherwise.

Furthermore, *Searching for Cloud Nine* is in no way endorsed, authorized or affiliated with any Cloud Computing organization or any of their similar subsidiaries. This book is not meant for promotional or advertising purposes.

Examples of people and other organizations that have or have not utilized Cloud Computing and its services are mentioned as case studies only. Any comments which could be deemed as negative or as criticism are completely unintentional on the author's part.

All information contained here is meant to be taken as a guideline. Every business is run in accordance with its own rules and regulations and the advice contained herein is mentioned in a neutral manner. It is understood that the reader claims responsibility for their own actions.

The author does not claim nor was any guarantee made regarding any success through this book. Therefore, he cannot be held responsible should any losses, risks, liability or damages that might be linked, directly or indirectly, related to the information contained within this book occur.

Contents

Chapter 1– Burden's Wheel

Our society rewards us if we build efficiencies and do it faster, better and cheaper.

In the distant past, companies were required to produce their own power to operate a number of tools and machines they used to produce their products.

In 1851, in a field beside an ironworks in upstate New York, Henry Burden built the largest and most powerful industrial water wheel in the world. At top speed, this 250 ton behemoth could produce 500 horsepower that generated enough electricity to power the tools in his factory through an intricate system of gears, belts and pulleys.

This invention made Burden a very powerful and wealthy man. But fast forward 50 years. By the early years of the 20th century, this once world class, cutting edge technology was abandoned, rusting in a field of overgrown grass.

The invention of "electricity as a utility" was a game changer. Companies no longer had to produce their own power. It very quickly became a competitive necessity to tap into the cheaper electrical grid than to produce it on your own.

The same technological revolution is happening today with cloud computing.

In October of 2005, one of the world's wealthiest tech billionaires wrote an urgent memo to his top engineers and most trusted management team. Along the banks of the Columbia River, a mysterious company known only as "Design, LLC" was constructing two massive 34,000 square foot, windowless warehouses on top of a fiber network code named "Project 2." The International Herald Tribune was quoted as saying it was "Looming like an information-age nuclear plant." The two companies were Microsoft and Google, the project was to build a cloud computing platform for their respective companies.

Many businesses are now adapting and opting to utilize the level of freedom that such a mode of storage introduces. While not just limited to storage place, cloud computing allows one to benefit from numerous practices that can make your business streamlined and efficient and you will never have to worry about losing your data.

If you are interested in learning more about the IT side of business, have a look at our IT book. It makes things simple and easy and you will walk away happy. Other than that, if you are interested in reading on for Cloud computing related information, you will be glad to know that you picked out the right book for that.

With the help of this book, we will take you on a cloud hunt for the perfect cloud computing program and what you should expect out of the whole deal. The search for cloud nine is by no means easy, but if you're willing to put invest your time and look in all the right places, then it can take you straight to Seventh Heaven for your business needs.

Gordon Moore - The Processor Prophet

In 1965, Gordon Moore, co-founder of Intel, predicted that processing power by using efficiencies of space and circuits would double every 18 months. This was affectionately named "Moore's Law."

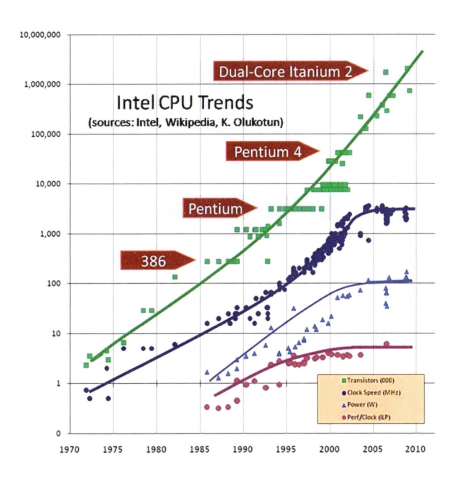

Just like the Burdens Wheel, "Moore's Law" is now deemed to be outdated and will not be true in the next two years. "The end of Moore's Law may ultimately be as much about economics as physics," says a DARPA director.

> *Despite the fact that 'Moore's Law' is outdated, computing power is still growing at a positive rate.*

So while this "law" is now outdated, computing power is still growing. While the speed of the single processor may top out due to physics, the number of processors we can tie together to accomplish a computing task is astounding. Some companies are prototyping 64 cores on one CPU. I like to think of a core as a lane on a highway with the speed limit equal to the processor speed.

Just think how far you could go with 64 lanes. Wow! This is all parallel processing which means that you have 64 lanes of data traffic all going at the same time. This type of processing power as well as virtualization has accelerated Cloud Computing to mainstream today.

What is cloud computing? Didn't you read the definition in the pages above? Oh, just skimming through, are you? Well, pretty much everyone knows you have to follow a rainbow in a vertical manner, find the cloud directly above it and you can use it for your computer. While you are at it, also try to find the leprechaun with the pot of gold. He is real greedy but he will totally fund your business for you if you catch him.

Just kidding! It is hard to find a rainbow. Anyway, cloud computing is a new trend in the IT department and has become rather widespread in practice and use. It is one of the safest ways to secure data. I would love to tell you more about it here but let's build up to it.

First, we will take a look at why the need for cloud computing was actually felt in the market and how various businesses are adapting to the assortment of technologies that are being introduced in the market.

Chapter 2 - How are Businesses Adapting to the Increase in Technology?

Every day, something new and technological gets pushed out into the market. Yay! But when you look at the rate that we are getting technological advancement shoved on us, you really start to wonder if there is a production machine that keeps on introducing all the technological advancements that we are experiencing.

> *Depending on your industry, it is possible to boost productivity and overall profitability of your business.*

Nonetheless, the rate at which technology is being launched in the market is not allowing us to have time to adapt to all its uses.

Smart phones and tablets might be fun for the average consumer but their full impact on businesses has yet to be gauged. Furthermore, various technological products have a variety of uses.

Depending on the industry your business focuses on, if you can get the right kind of business technology for yourself, you can end up boosting your productivity and your overall business profit.

Furthermore, while you might be shying away from using technology to your benefit, your competitors might have a different opinion. You will find that having technological advancements can really make you pull ahead in the race.

In the sprint for success, all the technological trends that are now being introduced can seriously boost you and aid you in reaching the finish line well ahead of the pack. Proof of that fact is the introduction of the Internet as a viable source of revenue for consumers and businesses.

However, technology has made more things possible then before and any business can offer online services or conduct business that is completely online in nature. Furthermore, there does seem to be a trend happening and the economy seems to be constantly utilizing the "adapt or die" methodology when it comes to changing with the times.

As a business, you either have to adapt and incorporate the newer concepts in your business or you have to close up shop since your business will have little chance of making it in the market. Ensuring that a business is flexible enough to be able to adapt to the changing times is one of the most important features that you need to incorporate into your business plan.

If you are scoffing at that fact, then just look at the way big established businesses have had to close up shop. Take a look at Kodak as an example. Considered to having revolutionized the photography industry as well as print development, Kodak had to close its doors owing to the fact that they were unable to 'shift' their business goals.

While the consumer market was gobbling up digital cameras like hot cakes, Kodak was unable to comprehend or make way for the change towards the utilization of digital cameras and the fact that film development was no longer a major part of photography and cameras.

By the time they did catch on to it and try to rectify the situation, it was too late and it was rather heart wrenching to see an organization as well established as Kodak having to bow out of the race in 2012.

However, you might believe that just not having the proper technology or not being flexible enough to make way for it is not very likely to cause your business to collapse. Yet it can raise the chances of such a thing happening. While just not having proper technology might not cripple your company, it can start a series of unfortunate events which might contribute towards pushing it over the red line.

And that is not the winning red line, that is the red line for bankruptcy, closure of business practices and the repo man coming to collect the office materials. That is not a very pretty red line to be near.

The Online Marketplace

While the trend was slow to take off, it has now become the norm that all businesses have a presence online. Even if you do not have a brick and mortar storefront, as long as you have an online presence, you are good to go.

The trend of online shopping and its successfulness can be seen with the establishment of renowned retailing giants like Amazon.com, eBay and others who are developed along the same lines. Furthermore, the online marketplace is not just limited to having a website on the internet anymore.

Social media websites like Facebook, Twitter, Linked In, Instagram, Pinterest, MySpace (is this still around☺) and other such similar platforms have begun to influence business trends, revenue generation and the kind of consumer feedback a business receives a lot.

An online presence has become a necessary requirement for businesses nowadays.

This in turn has started to affect the popularity of a business. For example, if you have followers on a business Twitter account or a Facebook business page, you are able to retain your consumers, get new consumers, share any new promotions, as well as interact with your target market as smoothly as possible.

On the other hand, a business that does not have an online presence, no website or social media page, will be looked on as extremely odd. If you are wondering why sales are slow, the market is not growing stagnant, your competitors have pretty much left you behind while they ran ahead.

Yes, you are an outcast and you can be sure that if your business was in high school, it would definitely be the kid that got picked last.

Furthermore, the everyday consumer has started becoming smarter and relies heavily on the Internet for information or to look up and search for products. Google is the largest search engine out there and anything you type in there can produce results based on the online matches it can make with your typed in search words. Not sure this is true? Well go on and type something as simple as 'the' or 'no' in the search box and you will see the search results it produces.

Owing to this increase in accessibility, an online presence has become an extreme necessity for most business practices. Furthermore, many businesses have also become iconic or known mostly owing to their online presence. Google, Microsoft and Amazon.com are some of the biggest examples of corporations that are renowned for their online presence.

Owing to the fact that consumers enjoy increased easy access to the internet, making everything available with the help of a few clicks or taps (if you're Googling on your smart phone or tablet), choosing not to give your business an online presence is pretty much boxing yourself in. You might as well mail yourself to Antarctica while you are sitting in that box.

Furthermore, businesses that do not have an online presence are seriously shooting themselves in the foot now. Socializing with the consumer through social media pages like Twitter, Facebook, the company website and the company blog makes an organization look more approachable.

On the other hand, if you do not have an online presence at all, you will be seen as the black sheep in the market. What are you hiding from the consumer? Why do you want to not interact with your target audience? C'mon! Quit making your business the Emo kid here and make it step out into the sunlight. All the cool kids are doing it but for once, there is no harm in this trend.

The Need for IT

Having an online presence can be like walking into a bar with swinging doors. You are going to push in the doors real hard, swagger in the bar, make intimidating eye contact with every one and let the doors swing close as you walk in.

However, there is also a chance that you will let the door go before you move and get smacked in the face with the swinging doors. That is what an online presence can do to you as well. Only you feel the pain right in the business. Or, right in the money maker.

'But,' you say, *'I'm one of the smart ones! I've got a good business going and I've got the proper social media and internet exposure. My target market is extremely fruitful, too.'* However, in spite of that, you are not experiencing much of a boost in your sales. So what's wrong?

Well for starters, having an online presence is not something as easy as throwing money at the screen and watching your website and social media happen. Incorporating an online presence with your business' identity should always be done with care. This also means that you will need to deploy a few measures that ensure that all functions on your website are completely streamlined and safe for users to access.

If you are not sure why you should do that, then go sit in a corner and think real hard about it. Now if you still have not gotten the point, we will spell it out for you. HACKERS!

Hackers belong to the deepest, darkest recesses of the Internet and are responsible for spawning viruses and other kinds or malware that wreak havoc with your online presence. With various viruses, some are able to manipulate your websites and even gain access to your servers.

Once there, all the sensitive data that you have will be at risk since many hackers take great pleasure in destroying your records and manipulating the data they find there. This has led to a lot of websites crashing, identity theft, internet fraud and basically turns your business into a playground for hackers.

Why they do it is not really something you should delve into since a lot of the time, they do it just because they can. On the other hand, if your website or blog or online application has a lot of virus or crashes a lot or has fallen prey to a hacker attack then it might not only hurt your business, it might also hurt your credibility in the market.

People will be more nervous about buying items from you or sharing their information on your forums. This is the swinging door that hits you right in the money maker if you are not careful.

When running a website that is meant for business, you should have the necessary means to safeguard it and that is where IT steps in. IT really stands for Information Technology, which we bet you already knew but are repeating it for the benefit of our novice readers. IT helps produce a secure infrastructure that allows you to safely use computers and other devices that can support the internet as a means of storing sensitive data.

Nonetheless, IT has served a huge part in changing the world around us. Not only was it a new way to work but the IT trend affected the whole world and brought the world's economy together with the help of its globalization effect. With the help of Information Technology, the online market has been able to expand to allow any consumer access to any retailer located anywhere in the world.

We are not exaggerating here. While it might be technically confusing, you will find that it is more than just globalizing the economy. It is cost effective, keeps hackers at bay and ensures that all your business functions and transactions go smoothly.

In fact, if you choose to outsource your IT needs to a company that provides in-house IT services, you will find that you have more time to focus on other core business needs which will help you expand your business and the target market.

An IT consultancy can basically be the nanny that you were searching for to look after your baby (your business) while you focus on setting up a trust fund for it to go to college. The young ones grow up so fast! Furthermore, having IT measures in place is like having safety gear on your business.

Having an online presence allows you to interact with a wider target audience

Moreover, having IT measures in place can help make your work easier for you. Your business will have more reliable services, you can offer your consumers excellence and you will be able to safeguard your sensitive data against hackers.

The IT measures you deploy are pretty much the bouncer at your party since not everyone is allowed access to the sensitive data that is stored within.

The Need for Proper Storage

Despite the availability of many kinds of technologies, there is one thing which is still deemed as a conundrum. This is storage of data. While our ancestors were just lucky enough to write on tablets of clay and papyrus, we have started utilizing a storage process in the form of paper bookkeeping.

As time went by, paper became a better substitute for clay since clay is sticky and gooey and had to be baked properly in order to preserve the writing. Paper storage meant having files, folders and big, bulky organizer shelves to keep those things organized.

Disadvantages of Paper Storage

Unfortunately, this storage method had a huge flaw. If kept sloppily then you could easily lose a document or misplace a file in the huge pile of paperwork. Furthermore, they were not disaster proof.

Disaster risk is needed for when anything manmade or naturally occurring happens. This could include fires, earthquakes, tornados, floods, etc. You name it, it can happen.

Unfortunately, moving the bulky organizer cabinets seems to have been a huge problem. Just imagine, your office is on fire. You have gotten all your employees out and all that is left are the precious, precious files.

You run inside and try to drag out your organizer cabinets. You also only have 90 seconds to do that. Oh look, you just made it to the stairs! You are out of time. You are dead and your files are destroyed.

Now look at it like this: A flood afflicts your city and comes pouring and roaring up to the office. Your office gets submerged in the water. Before you swim away, you turn to save your precious paperwork but realize that half of it is just pulpy mush and half of it just sunk to the bottom in your organizer cabinets. You swim to safety, wait till the waters recede and then go back to your office.

There is a 99% chance that your paperwork is utterly destroyed and a 1% chance that you will be able to find your office among all the rubble, mud, dirt and debris a flood brings.

Despite the impracticality of the whole process of storing data in paper and filing cabinets, this obsolete method is still widely practiced in numerous businesses. This in turn has put many businesses in jeopardy when they are missing important documents.

Furthermore, paper documentation can be more easily destroyed and manipulated which can also lead to a lot of theft, embezzlement and other serious complications.

In order to make a safer storage system, the use of floppy disks, CDs, hard disks, hard drives and external hard drives became pretty commonplace. Unfortunately, these are also only as good as the hard drive they are stored on.

Once a hard drive or other memory device is damaged, the data can then be irretrievable. In the quest to finding a proper storage device, people started looking into better IT resources in order to make storage easier to retrieve and easier to keep secure.

Owing to this need for proper storage, many people have started turning towards the new kinds of technologies that are being introduced in the market. Although utilization of one is not quite enough, you will be able to make a significant increase in your business' storage properties.

For those of us who started out in the world where many of the technological advancements had been made, an office that depends entirely on paper data seems highly inefficient and quite unheard of in today's business management practices. But rest assured, it used to one of the best and most widely used ways to keep files. That is, until it got outdated.

With newer technological advancements being introduced for storage, paper finally lost its hold as being the only universal means of sharing information. However, the newer technologies, while being rather eco-friendly did come with their own set of unique problems.

Chapter 3 - New Technologies

In order to make a safer storage system, a lot of new technology was introduced into business practices. Businesses as a whole became more streamlined in practice but this was something that did not happen overnight.

Little leaps and bounds introduced newer innovations such as the typewriter, which then got replaced by the word processor and finally got replaced by the computer.

The Birth of Newer Technologies

The traditional telephone got a makeover with smoother keys and sleeker interface which was then turned into a cell phone. With slow advancements like these, the possibility of having online, virtual conversations through the internet was also looked into. Once the computer was introduced, it was only then that advancements to improve the paper storage system were made since efforts were, and still are, being made to greatly streamline various manually conducted business processes and transactions.

This step introduced and incorporated the use of floppy disks, CDs, hard disks, hard drives and external hard drives as pretty commonplace in business. As a new means of storage, it brought about a great transformation in the storage of data. However, there was one thing which made the utilization of hard disks and external hard drives really cumbersome for offices.

The Disadvantages of the New Storage

The servers and other storage items had to be housed on the office premises. Not only did this increase the need to get wider and larger offices in order to accommodate and provide enough storage space, they also required huge bulky servers and constant server support and updates to look after 1GB of memory.

It was also so expensive that mainly the rich and well established businesses were the ones that could take advantage of these methods while the other businesses had to make do without them until the modern day, pocket friendly and supersized version were released.

Not only was this a slow trend to catch on for businesses and organizations, it is still not always doable due to the cost factor that starts coming in. This is one of the main reasons why many offices and business are still largely utilizing paper data practices to this day.

Unfortunately, the introduction of the hard drive as a viable storage device was also not a very great success since these devices are also not very good or safe for storage. This is because your storage options are only as good as the hard drive they are stored on. Once a hard drive or other memory device is damaged, the data can be considered irretrievable based on the kind of damage it undergoes.

Let us take a look at them in disastrous situations for a fair comparison. Say your office is on fire. All of your employees are out except for your precious, precious files.

You run inside and you just have 90 seconds to get all the hard drives out. Oh look! You just made it to the stairs that lead to the floor where they are stored! You are out of time. You are dead. You died because your office was too huge.

Well, it is not very probable that this will happen but, if it did, your files might have a 10% chance of retrieval and that depends largely on the fact that the fire did not destroy them or the fire department did not hose down you electronic hardware. But hey, a 10% chance. You will take it. Water and electronics do not mix well. Either your computers short circuit or your hard disks go "kapoowie" or the water just acts like a good conductor of electrical currents and fries everything in its path. Since natural physics always trumps in such situations, option two is more likely to happen. That is terrible! I can't even place any numbers on that.

Because of this need for proper storage, many people have started turning towards even newer kinds of technologies that are being introduced in the market. Although utilization of one is not quite enough, you will be able to make a significant increase in your business' storage properties.

In the quest to finding a proper storage device, people started looking into better IT resources in order to make storage easier to retrieve and easier to keep secure. But enough of the technicalities.

Chapter 4 - What is Fluffy and White Yet Holds Massive Amounts of Information?

This is where you guess ☺

Okay, those are really good guesses.

The Actual Answer

A Cloud!

Well, before you get mad at me, we meant to say an IT Cloud. It is white and fluffy like a cloud and holds massive amounts of information!

The Internet in a Nutshell

Once upon a time, a long time ago, we started the process of computerizing businesses operations and transactions. During that time, someone had the brilliant idea to hook up the computers to the Internet! Huzzah! Everyone shouted with joy for it really was very smart of them to have such a thought. The Internet at that time was already known as a hub of information that one could easily share data through.

Created in the 1950s, the earliest version of the Internet was further refined over the decades into what we know today. The modernization of the Internet has carried on through the ages with the modernization of our technology and, as yet, we do not think we can say that we are fully done tinkering around with it.

But we digress. If we start looking at the history of the Internet, it is far too vast to summarize, even in the pages of this book, let alone the chapter so we will skip forward a couple of decades and come to the point when emails and websites were introduced.

People could finally talk and communicate with each other without having to leave their cubicle and business could be easily conducted without needing someone to get on a plane and fly halfway around the world just for a signature. (Even if it was kind of awesome.)

This streamlined way of sharing information was restricted and reserved to certain connections and maintained through a server which not everyone could gain access to. This secret server also harbored many business secrets, data and information which slowly but surely led to the birth of 'the Hacker.'

Well, not exactly the birth of the hacker since the Internet is certainly not responsible for siring that unruly child but, it turns out that people love to disrupt organizations and disrupt the computing system every now and then. Surprisingly, it turned out that hackers were remarkably adept at these kinds of things.

Enter the Cloud

Inheriting much of their skills from the Internet, malwares developed by hackers have adapted in accordance with the various new programs that are now being introduced. However, the IT Cloud was something that helped to answer all these dilemmas with one fell swoop.

It provided a sweet, secure storage area and instantly smoothed out the IT requirements that many businesses have. Fondly getting the nickname of the cloud, it is not really shaped as a cloud. It is purely a virtual connection which allows you to store data, access files, etc.

While it was just a budding idea, a cloud-ling if you will, there was probably a lot of doubt about it being of much help with IT becoming such a huge and necessary part of the business industry. But I digress again. So where does the IT cloud come into all of this? Why dearest impatient reader, the Cloud is the hero in this fairy tale! And you are the princess.

The Cloud came as a solution in the business market, particularly for small businesses and other mid-sized businesses who often fell victim to the malware attacks of various hackers.

The virtual storage system that Cloud Computing presented them allowed them to boost their productivity and indulge more freely in money making activities. They were also able to promote innovation and indulge in collaborative work with other businesses. Something they had been unable to accomplish online before.

Furthermore, over time, the cloud has been refined and looks nothing like its earlier version. Not only have its limitations as a cloud storage device expanded but it is now also able to support a vast number of platforms.

Through these, it becomes possible to share information and to indulge in hosting activities such as hosting applications, web based phone calling systems, email services, data storage and much more. These hosting activities are the ones that have earned the cloud its title of cloud hosting or cloud computing trend in the market. However, you must probably think that cloud computing is a term that gets thrown around a bit too much these days.

There is a reason why Cloud Computing is so popular in the market.

Chapter 5 - Why is Cloud Computing Such a BIG Thing?

You must realize that everyone, including this book, is talking a lot about the IT cloud. Any work done through that method is known as cloud computing. We will break it down for you. The Cloud was already being utilized as a metaphor for the Internet. Yeah, we did not come up with that so don't look at us for an explanation. Anyway, when you throw in the word *'Computing'* with Cloud, you get a whole different concept.

Understanding Cloud Computing

Cloud computing is a pretty big thing in the market with people advocating its use like it is the only thing in the world that could ever help businesses exist. However, that is not the truth. You know that and I know that and I am sure we are both comfortable enough with each other now to be honest about it. But before we get started on why cloud computing is such a big thing, we will take a look under the hood to find out what the cloud really is.

If you search for cloud computing, you will find it defined as not having a definite term. It is a slang phrase which caught on and became the name for what you would call a network of computers. Cloud Computing is actually a type of computing which is similar to grid computing.

Virtual servers are utilized in order to store, create backups and work as the file cabinet for your office. Even though it is still believed to be in its initial stages of development, there is still belief that there is much to be done. It spent a lot of its early life just floating around the internet as a droplet of an idea and finally took shape.

Nonetheless, even in its initial stages, the introduction of the cloud has pretty much revolutionized the IT industry in quite a different manner, allowing increased productivity and business security.

We will put it in simpler terms. You no longer have to worry about hackers, disasters that have you battling through fires or swimming through cold expanses of water. All you have to do is place your files on this virtual server and watch it float away to safety. You can also get your files back any time you want.

How Does Cloud Computing Work

Cloud computing works in a brilliant manner which allows you to enjoy increased productivity as well as enjoy business security. We will put it in simpler terms.

However, let us take a look under the dashboard like we promised. Now, to make this easy for you, we are going to divide it into two parts. We will call them the front end of the cloud and the back end of the cloud.

Front End of the Cloud

Now remember, the cloud really does not have a proper front end but we are breaking the IT Cloud into two parts in order to make it easier to discuss the way it works. Now in the front end of the cloud, you will find the user. That is you and this is your end of the computer.

Over there, we find you working away or reading something or you are just pretending to look busy. Anyway, this is the user's end and this end also refers to the usage of any application that allows you to sign into the cloud. This could also include your browser system or something that you downloaded onto your computer.

Back End of the Cloud

Now the back end of the cloud houses all the different servers, data storage systems and other servers that complete the Cloud Storage system. Now to successfully utilize cloud computing services, you need to utilize both these items together.

Theoretically, a cloud computing system is supposed to be able to support almost any kind of computer program that you have in mind whether it's just a basic data processing program or a videogame. Usually, in cloud computing, each application gets its own dedicated server to ensure optimum performance for anyone on the front end.

When utilizing cloud computing methods, there is a central server which serves as the administrator and is used to basically monitor the system, keep track of the clients and the demands that are being made on the system. With this administrator, everything runs smoothly and ensures that all the storage is being properly utilized.

Ta da! That is the cloud and this is why it is so big. It allows many businesses the chance to work easily, efficiently and quickly.

That's Cool and All But Why Should I Get One?

As a person running a business, getting Cloud Computing for your business is picking a baby sitter for your baby while you go out for an evening.

Luckily, you are not the only one who has raised that question. Many businesses have started to pay attention to the cloud and the question on everyone's lips is, "What is the Cloud? What can it do for my business?"

We will give you a brief overview of why you should get a Cloud Computing System. This also helps to highlight why Cloud Computing is so BIG in the market.

Life Before Cloud Computing

Before Cloud Computing came into the market, the traditional applications for business that one had to apply were extremely expensive and tedious. They were also complicated to apply. There had to be a special team of experts required who had to look after all that software, update it, test it and secure it as well as be in charge of ensuring that things run smoothly.

Therefore, it was pretty daunting to get the required software and hardware for a business. Nonetheless, with the advent of Cloud Computing and its success so far, many small businesses have started to utilize Cloud Computing services as the perfect complement for their business.

This has even helped many businesses finally make the leap and take the next step which had previously eluded them. In other words, cloud computing proved to be the stepping stone they needed to jump across the river. Do not try that with an actual cloud though! We heard it is not actually white and fluffy.

Tailor Made for Small and Medium Businesses

When you are starting out, you have a set budget and you cannot afford the luxury of investing in all the software and hardware that is usually required to maintain them. It was also very hard to find hardware and software that corresponds with your budget. This was one reason why only large businesses used to utilize computer storage practices.

However, with the help of Cloud Computing you no longer have to worry about that. Not only have websites become more pocket friendly but you will also be able to cut down almost half of the requirements that are needed for your business. For small businesses, cloud computing is the best option in order to streamline their business, cut down their expenses and focus on other business needs at the same time.

Computing Fast and Easy

With the help of Cloud Computing, all those daunting requirements are immediately cut down. You no longer have to worry about getting a team to maintain all the hardware and software.

Cloud computing solutions dramatically help you to cut down and eliminate all those countless headaches and responsibilities. You can now let a vendor deal with it by paying them to look after your IT through their cloud computing expertise.

Furthermore, if you have developed a cloud based app, you can dramatically cut down on your expenses by allowing your employees to sign into a virtual office. All they will have to do is open their browser, log into the cloud app and send their work directly to your database.

You also no longer have to worry about getting security. Cloud solutions are offered with rigorous and confirmed security and storage services. You do not even have to invest in extra money for security since it comes as part of the package that you will be getting from your Cloud Computing services provider.

Testing the Storage Capability of the Cloud

Now we are going to take look at the ability for the cloud to function as a good storage device. We have already put paper storage, the hard disks, hard drives and servers that are used as storage through this test so you can expect we will be doing the same for Cloud storage.

It is only fair to the other storage devices and will further solidify our point regarding its super awesome storage purposes.

So let us imagine that you use Cloud Computing as a means of storage. You have all your precious data stored there and it is easily accessible to you from any device that supports the Internet.

So back to the fire. All your employees are out and all that is left are the precious, precious files.

Since you use Cloud Storage, you finally run outside after making sure that everyone was out. Look at that! You did not die, you little trooper, you! Despite the fact that your office burned down and any paper files you might have had inside your office, if you used Cloud Storage, you do not have to worry about them. Just relax, roast a marshmallow and wait for the fire engines to arrive.

Once you have access to a computer or a laptop or even a tablet or a smart phone, you can easily log into your cloud storage system and gain admission to your files.

Mind you, the duty of keeping those records updated does fall on you but if you have been keeping your Cloud Storage files updated, you can easily expect a 60% chance of full retrieval of your files, a 30% chance of semi-retrieval of your files and a 10% chance that you completely forgot to utilize your cloud after you got it.

Well, hey! Look at that. The numbers are great! 60% is awesome compared to the 0% promised by paper storage and the possible 10% offered by hard disks. Now let us put the Cloud in a flood.

Water, water everywhere and more than a drop to spare. So just keep swimming like Dory taught you and you will be away from the nasty electronics and the electric current that might roast you like that marshmallow you were roasting in the fire. Wow! That is not a nice picture. Ignore us and keep swimming. Good! Wow, you have become a great swimmer. You have survived three floods now!

Now when you have reached a safe distance, you turn back and see your hard disk go down (No not Disco

Down). Once that happens and you have a chance to access a laptop, a computer, a smart phone or a tablet, you can easily log into your Cloud Storage system and gain access to your files.

If you have pretty much kept your data updated, I am afraid we can only offer you the same figures again. You can easily expect a 60% chance of full retrieval of your files, a 30% chance of semi-retrieval of your files and a 10% chance that you completely forgot to utilize your cloud after you got it.

Now this is awesome! Cloud Computing allows you to enjoy your business as securely as possible. However, if you have already established a business without getting Cloud computing services then we do applaud you for your audacity. While it is okay to start off without getting cloud services, there are certain benefits that it provides.

Furthermore, if you feel like you are stuck, Cloud computing services can often give you that extra benefit and security that you need in order to progress to the next step in your business.

Chapter 6 - When Do You Need to Get a Cloud?

Now this is the million dollar question. At what point does one think he or she should get Cloud Computing services for their business? How does one even learn that their business requires Cloud computing services?

We are not going to sit here and pin point the exact the time that you need to switch your software to the Cloud. We are going to believe that you are a good enough judge of that.

Like we said, getting an online presence is not as easy as throwing money on the screen and watching your website and social media happen. However, what we will do is give you simple tips or bullet points that you can see. These can help you evaluate your business needs and gauge the answer.

Just think of us as your very own Yoda only we are more bookish and small. Plus, we talk more coherently. Only Yoda can pull off that backwards grammar! Now that we have established our mentor status, we should get back to subject. Even if you do not feel like you need the whole set up, utilizing Cloud Services allows you to safe guard your website and everything you have in there. If you are running an online store, getting Cloud Computing Services may be an absolute necessity for you as your business expands.

Let us not leave you scratching your head anymore. Here is a list of reasons which indicate that you should get Cloud Computing services.

Looking to Incorporate Scalability and Elasticity

When you are running a fledgling business, you are going to need computing services. Everyone understands that but you have to get computing services that match your business' growing needs. A growing business is like a growing child. You need to make sure that when they get into their teenage years you are able to accommodate their growth.

You would not make a teenager wear their childhood clothes and so, you do not give your business the same computing services that it had when it was just in its initial stages. However, you cannot give them something oversized either since your business will just lie there, rolling around in the extra rolls of computing.

On the other hand, Cloud Computing services are like the jumpers your business never had. The elasticity of Cloud computing allows it to be utilized in accordance to your needs. The flexibility of the Cloud automatically makes scalability possible for your business. This means that, as your business grows, your cloud grows with you, too. When you are looking to expand, the cloud is the perfect jumper for you!

Saving Money While Maintaining A Limited Budget Is No Problem

When you are setting up a business or running a business, you often realize that everything is allotted a certain budget. Go even one step over that red line and you find that everything gets pulled out of place.

The ensuing disorder can often be responsible for harming or crippling the business in such a manner that it can make growth of the business impossible. Not only that, but you might find that navigating a crippled business is like sailing a ship with a broken hull.

You will either lose direction or come to a standstill or just start sinking slowly into the murky business waters.

Therefore, you will need computing services that will not punch a hole in your wallet. The Cloud is the one you want. Coming forth like a fluffy warrior, the Cloud will provide you with all the cover, backup and support you will need to make navigating through the stormy waters easy for you.

The Cloud is the magnificent sail to your ship and will help you navigate through all the rough patches in the ocean. This is because you can pay for the Cloud Services you utilize in accordance with your consumption ratio. That means that if only 30% of your business is dependent on the Cloud, you will only have to pay for that 30%.

This is not only a very pocket friendly alternative for many people but it has allowed many businesses to grow in a safe and secure manner. When you are looking to utilize the proper Cloud computing services, this option allows you to avail that easily without putting any additional strain on top of your already restricted budget.

With such freedom and the increasing development of the Cloud, businesses and people alike can easily store their work and safeguard their data as easily and as inexpensively as possible.

Chapter 7 - The Search for the Perfect Cloud – Cloud Nine. Where Are You?

So now that you know what the Cloud is, you must be all aquiver to find your special Cloud Nine. It sounds so good that you would love to have it right now. You will have to look among a field of clouds to pick out the one which is the best. When you find your cloud, you will know it is yours. It will wrap itself around your business and hug it till it produces money.

That is how much a good cloud computing system is capable of doing. Yet when you look for it, the task is not easy. All the other clouds are just as snowy white, if not whiter and prettier and, at first, all clouds will seem the same to you.

Yet, you will find that over time, it is possible to utilize it for an array of tasks. First, you need to find your perfect Cloud Nine. This is the ultimate cloud that you will feel is tailor made for your business.

So when you are out in the cloud market, the following are some of the ways that you can try to find your special Cloud Nine.

What Kinds of Clouds are Available in the Market?

I am glad you asked. You are getting really good at asking questions and making us come up with an answer! We have trained you well it seems. You will find that there are three kinds of clouds available in the market so far. They are known as the public cloud, the private cloud and the hybrid cloud.

Now before you start picking one or the other, it is best to take a closer look at the kind of services they provide.

While you might think that you can easily make an educated guess based on the kind of services they provide by their names alone, we are going to urge you to a look under the hood with us before you make your choice.

The Public Cloud

Built on an external platform, a public cloud is mainly run by the cloud service provider. With the help of this off-site cloud option, users get to utilize their own cloud as long as it is within that particular shared infrastructure. This is also the most basic kind of cloud available and it is based on the standard model of Cloud Computing.

The provider is in charge of providing all the resources to your system which could range from just security services to maintenance of your Cloud as well.

Owing to the fact that it is largely managed by other outside companies that offer particular services and specialize in offering cloud services to a large variety of customers, the public cloud can be the perfect companion for your business if your core focus is utilizing the latest technology in a cost effective manner that also ensures elasticity of use.

In most cases, the provider makes services available to the company over the public medium of the internet. Often times, such cloud services are pocket friendly or offered free to the public, like DropBox or the Windows Azure Services Platform. Some of their cloud features are free for the public to utilize but they can pay to get more space. This successfully incorporates the pay-per-usage aspect of the cloud and allows a larger degree of scalability for businesses and other users as well.

The Private Cloud

If you are looking for something more business oriented, you can opt for the private cloud. This shy elusive cloud will be your very own and you can easily install it as a compliment for your business' firewall. Also known as the corporate cloud, these kinds of clouds are more highly protected and require special passwords and other authorization keys before the information stored within cloud can be accessed by a third party.

However, private clouds are more widely utilized as storage spaces for many offices that use cloud services as a means of keeping a safe and secure backup. They are also preferred by businesses who want to have a wider measure of control on their data, which is not always the case when a third party is in charge of handling the cloud computing needs of a business.

However, this limits that maintenance of the Cloud to the IT department only and there is little to no involvement of the public or the service providers. All solutions and services are managed by the IT department and done so privately within the office's infrastructure.

For businesses that have an in-house data center, the private cloud is the most obvious choice since it enables the management of all IT hardware on a local level while it enables the employees to gain access to the database remotely. The private cloud has become the norm for most business practices nowadays and is indeed among the most common type of cloud on the market.

The Hybrid Cloud

This bold little cloud offers the best of both worlds. It is a mix of the public cloud and the private cloud and incorporates features from both as well. Many businesses often opt for this customized version owing to the advanced flexibility it offers. In other words, when utilizing a hybrid cloud, you will be able to manage some features in-house and will have to outsource for some of your Cloud related needs.

In particular, businesses that need to interact with the public to some degree will benefit from these kinds of clouds since it enables them to interact with their target audience and ensure that all the data is safeguarded.

When opting for a hybrid cloud, it must be understood that such clouds are created specifically, so there might be some discrepancies in the system. The main aim with the hybrid cloud will be to smooth out the creases and avoid too much change. The larger the design differences within the Cloud's environment, the less effective the Cloud will be.

The hybrid cloud also allows businesses to take advantage of the cost effectiveness and scalability of the public cloud while allowing you to enjoy the privacy and security of the private cloud.

The best way to decide if you need a hybrid cloud is to start with either a public cloud or a private one and then incorporate the other cloud environment. With a strong base that supports your business, you will be able to have a Cloud which strongly correlates with and meets all your business needs. So far, only these three types of clouds are available on the market.

However, further research is still being conducted to get the most out of the Cloud and see to what heights it is possible to push the little Cloud. So keep your eyes and ears open. It may be possible that in a year or two we may stumble across more efficient and elaborate clouds. Remember, when you are searching for the perfect cloud, the sky's the limit!

Chapter 8 - Help! I'm Scared/Confused/Worried or Have Doubts and Fears About the Cloud!

So now you are one step closer to getting your very own Cloud. Good for you! But what's that? You are not sure if you should go for it? Well, hey, there is no need to be scared or confused or worried about the Cloud.

We have spent a lot of time learning the benefits and services that a cloud can provide. How about you take some time to explore your Cloud before you choose to get it?

Many people often do not have much of an idea about what a Cloud is and what it can often do for their business. This generally leads to misconceptions about the Cloud. However, we will take the time to address a few of those misunderstandings and show you that the little Cloud is really a white knight in some rather fluffy armor.

"I Need a Little Cloud, Not a Big One"

Many people are under the assumption that clouds come in all forms and sizes. It is probably as simple as when you go to buy a small bottle of ketchup from the store. No, dear reader, no. First of all, the Cloud is pretty much limitless. No one is aware of how big the Cloud can actually be. Well actually, some people are.

When you are looking at Cloud based services like Facebook, Google and Amazon, it is really, really hard to make an educated guess. We would like to say 1 gajillion but that's not really an IT term or a proper number so our answer has been refused to be included in the records. However, according to a super huge info graph, the correct answer is one exabyte.

If you do not get the term then not to worry, the info graph took the time to break it down and explain it in credible terms. One exabyte is capable of storing around 1 billion floppy disks, 2.68 million flash drives and 1.6 billion compact discs within it.

However, the proper thing to do here is to utilize the pay-per-usage plan. This will give you access to a little bit of the Cloud and allow you to increase your usage as you grow.

"Our Business Really Does Not Need Cloud Based Solutions"

If you will remember, we mentioned how businesses who fail to adapt can hit the red line of bankruptcy with all the subtle grace of the Titanic crashing into the iceberg.

Cloud solutions are meant for everyone. It is a platform to interact with your consumers at a remote office that allows your employees to log on and work from anywhere in the world or even as a credible means of storage. Do not be scared of the Cloud. It is totally safe and harmless and is becoming a big part of everyday life. And it is cute and fluffy...or manly and fluffy!

Do not turn your back on it. You might also be utilizing easy access clouds like DropBox, etc., which might make you think that you do not need specific Cloud based solutions. However, something like a private or a hybrid Cloud is more suited for a business.

You do not want to foolishly do that to your business. A public cloud is good as a temporary solution only since the process of deploying it is still fairly new. Be sure to change to a hybrid cloud if you want but do not abandon the cloud so callously.

"The Cloud Is Just a Passing Fancy"

Well, no. The cloud was not and is not a passing fancy for anyone. Whether you like it or not, the cloud is here to stay and the best thing to do is to just give in. Do not fight it. It is good for you.

Now that might have sounded creepy but we do mean it, the Cloud is good for you and it is flexible enough to accommodate almost any business' needs. Whether you choose a public, private or hybrid cloud, you will find that your business takes to it like a duck to water or a fish to water, or a dog to water. You know what we mean.

Another thing that you should keep in mind is that even though using the public Cloud is relatively new, the concept of using a private Cloud is not that new at all. In fact, it has its roots firmly planted in old IT practices that were being utilized ages ago. A rose by any other name would smell as sweet and the Cloud with any other name would continue to streamline your business needs. Now that is sweet.

Nonetheless, the Cloud has actually been around for some time and many businesses have successfully utilized the private Cloud for their business practices. Thousands of organizations actually deploy the private Cloud on a worldwide scale. That is why choosing to believe that it might be just a passing fancy could turn out to be the worst thing for you to do. That is like saying that dragons do not exist!

There is strong proof that shows that Cloud usage is growing, owing to the fact that there are three versions of it now where once there was only one. The terminology for the Cloud might have changed over time but the basic premise on which all private Clouds are founded and deployed on is the same. That is why it is wiser to be ready for the change rather than try to resist it.

”The Cloud Is Not In Our IT Road Map”

Many offices that choose to resist the Cloud deliberately ignore the Cloud through this excuse. Do not be that guy. However, it is completely understandable that you might think that your hardware and software is not equipped for the Cloud. In such cases, the best option is to give the Cloud a test run. Use a public Cloud and use it on a pay-per-usage basis. This way, you will be able to give it a proper test without having to spend too much money on the Cloud.

The Cloud is something which a lot of businesses have had to make room for so choosing not to acknowledge it by claiming that it has no room in your IT roadmap cannot just be damaging to the Cloud's feelings, it is also a move which can harm your business and the growth of its infrastructure.

You should take the time to realize that IT and Cloud are two things which are constantly changing and adapting according to the latest shifts and trends in the economy. The proper utilization of both elements can ensure optimum results for your business and that will become more than clear in the manner your business runs and conducts work with their help.

However, the Cloud is seen as the right hand man of IT. If it is not there to facilitate and support your IT practices, you might find that your business might become precariously balanced on three legs like a lop-sided table which could go over anytime.

In the end, the best thing to do would be to get in touch with a Cloud consultant or a Cloud expert. There is no harm in consulting the professionals and finding out if your business really requires the Cloud or not.

Just do not dismiss it on its own since that could end disastrously for your business. We have already mentioned the downfall of businesses that are too slow to adapt and we are pretty sure that you are not interested in re-hearing the gory details here again.

"Cloud Does Not Offer Enough Security"

This is one of the biggest reasons why people will not opt for the cloud. They want their data to be in a place which is as snug and safe and inaccessible as an eagle's nest on a cliff. However, a large portion of people feel that the Cloud cannot offer enough security to safely provide such a nest for their precious and beloved data.

On the other hand, it is becoming clear that people are looking at the Cloud from the wrong side of the telescope. Of course the Cloud will look weird but turn it around and look at it through the right end and you will realize that the question is not, *"How much security does the cloud offer?"* It is, *"How can one ensure that all the data is safeguarded with the right security measures in the Cloud?"*

When you are getting yourself a Cloud, be sure to ask your Cloud service provider this question. The proper Cloud has measures deployed that complement its cloudiness.

If you are using a public Cloud, you will find that security measures deployed by them are very high. Since public Clouds ensure more interaction with the public, the security measures in place will generally be higher and tighter to avoid any attacks from malwares.

When you are using a private Cloud, you will find that the security measures deployed are more in compliance with your organization's rules and regulations as well as its own security measures. This allows for security measures to be completely in the hands of the client instead of the service provider.

Therefore if the Cloud does not have the right security measures, it will be important to scrutinize your security measures and reevaluate them as soon as possible instead of blaming the poor little Cloud for not performing its duties properly.

"I Wo Not Have Enough Control"

First of all, that is not a very healthy thing, you know? Wanting to control everything around you can make you go crazy because, surprisingly, the world does not work that way. On the other hand, if you are still looking to get a Cloud which you can have under your complete control, then the best option for you would be the private Cloud.

The public Cloud is, in many ways, largely controlled by the service provider. You can start out with a private Cloud and later on move toward a hybrid Cloud if you feel your Cloud is too big for you. However, when you opt for a private Cloud, you still get to enjoy a few benefits of the public Cloud, such as its scalability, elasticity and the ability for self-service.

A private Cloud will also be under the management or controlled directly by your IT department. This means that they will be in charge of deploying the Cloud correctly and also ensuring that it is implemented seamlessly with any hardware, operating system, network, etc.

The duty to ensure that all the security measures are properly deployed for the Cloud also falls on to your IT team. No doubt, deploying a private Cloud will mean changes in your data center but you do not have to worry.

Those changes are completely under the control of your
IT manager.

As the Cloud expands, you might end up with a lot more on your hands, which might be too much for the IT team to handle. If that happens and if you are finally tired of being so controlling, you can shift towards a more hybrid Cloud and outsource some of the work to someone else.

As you can see, if you are avoiding the Cloud since you are thinking it will not offer you enough control, then you should definitely reconsider. The Cloud is considered flexible for a reason and this is one of the reasons. Well, speaking of flexible, here is another reason.

"The Cloud Will Not Integrate Well with Our Other Systems"

The Cloud's flexibility is not something which can be tested like pulling a rubber band to its utmost limit.

It is measured by seeing how quickly, smoothly and seamlessly it can adapt to your hardware and software as well as the rules, regulations and other business practices that you might follow.

On the other hand, with all the software and hardware that is being sold on the market, they are being made in a manner so as to incorporate the Cloud in them as seamlessly as possible should you decide to integrate the Cloud with your business anytime in the future.

Furthermore, it is now a simple matter to utilize your existing hardware with the right kind of software which will enable it to build your very own private Cloud. In the marketplace, you will now find all the correct tools that will help you to securely bridge your hardware and software with your private Cloud and the other applications of your business.

However, this can, at times, get too confusing to handle. Especially if your IT team is as clueless as you are. Do not feel bad about it. Ignorance is bliss sometimes and it is very easy to perceive the Cloud as a relatively alien being which wants to gain entrance to your office. Do not be terrified and panic, though. The easiest thing to do is to find a Cloud consultant who understands the Cloud.

With this person's help, you can devise a road map or, more likely, a Cloud strategy which will help to work out what you can do with the existing database and how you can plan out your future growth along with the Cloud.

Do not be shy about asking for help. It is better to stop and ask for directions rather than getting lost and having to pay a large sum of money to rectify your mistakes. This will also help to plan and strategize properly for your business' growth ahead.

Chapter 9 - So You Have Your Cloud Nine and Want to Grow Big Enough to Make It Rain

Now that we have helped clear away all your misconceptions about the poor Cloud, let us look at what you can do to grow big with your very own Cloud Nine. While you were busy reading our tips, we took the liberty of laying a silver net over the rainbow and catching a beautiful, fluffy Cloud for you. Or, Cloud Nine, as we like to call it.

Now you want your Cloud Nine to grow from just a simple cloudling into one of those mighty thunder clouds. As your business grows, the Cloud Nine will grow with you. Do you want to make it rain? Well, we have just the right set of tips with which you can bend Cloud Nine to your will!

Create a List of Requirements for Your Cloud

First of all, start listing your IT related business requirements. Be as honest about them as you possibly can. This will help you to realize how big you can push your Cloud Nine to become. If you do not have a lot of IT related requirements then it is rather improbable that your Cloud will grow big anytime soon.

Be brutally honest with yourself. Like, lay your soul on the chopping board and chop away any extra pieces. One of the things I tell my customers is they need to tell me when my baby is ugly. That way, I am aware and can fix it.

There is nothing wrong with dreaming big but that should not make you needlessly invest in items that you will one day hope to utilize. If your IT needs only extend to the utilization of your account and consumer interaction, then let your Cloud Nine take over these areas. As you expand, your Cloud will expand to accommodate more and more portions of your business' IT needs. But if you are not at that point yet, do not add it to your list of requirements. Do not worry though, the Cloud is flexible enough to accommodate any additions that you may add later to your business.

Try It Before You Buy It

We all know how awesome Cloud Services are. We have been harping on and on about them the whole book, right? But in the end, when it comes down to it, you might have to do a little more searching for Cloud Nine before you get the one you like and the one you want. There is no harm in trying and testing the Cloud of your choice.

When you are testing out the Cloud, look for the following services from your Cloud Service provider:

- The service provider adheres to formal policies that respect your confidentiality and your privacy and the service provider is required to seek your permission in order to gain access to your data if any support is needed.
- Ensure that the service provider has or will incorporate data logs. These will keep track of who accessed your data and when that data was accessed by a person. The reviewing of the data logs will, of course, fall to you.

- Prior notification will be given if any software updates are being incorporated that may or may not affect the nature of the data you have stored.

- The service provider cannot use your data for any marketing related purposes or any promotional activities. If they do wish to do such a thing, it should be mandatory for them to contact you and request your permission first.

- The service provider clearly outlines the methods they utilize in order to provide you with the best possible security for your data. These should be clearly outlined in their contract agreement. Be sure to go through this contract agreement as thoroughly as you can. The statement should also include the measures that can be taken in case the agreement is violated.

If you think this is being picky and you are too shy to tell your business provider what you want, then remember that you are in charge of the data that you have and it is your right to ascertain the best possible means to ensure its safety. In the end, if any breach of security does take place, it is not the service provider who will suffer the wrath of your clients, it is you and your business.

So take the time to make your choice. Be picky and be careful when you are picking a Cloud Service provider. Furthermore, there are plenty of Cloud Services available in the market so if you do not find the best one straight away, it is not going to be the end of the world for you. You can easily try and test different Clouds from different Cloud service providers.

Many Cloud Services also allow you to go through a trial period. If you are not satisfied by the end of it, you can easily change your Cloud service provider. There is not much of a difference in them but, in the end, you are looking for the one that seems to integrate with your business as smoothly as possible.

Making IT Someone Else's Headache

When you are just a small business, you probably do not have enough resources to pay for or house an IT team that would be at your beck and call to look after your IT related needs.

On the other hand, by getting Cloud services from a Cloud service provider, you are able to effectively outsource your IT requirements to your Cloud service provider. However, do not assume that the Cloud service provider will immediately look after all your business' IT related needs. They will only cover your Cloud related needs but, even those are a substantial amount.

They will look after all your Cloud computing software related needs and also ensure that the cloud is secure and kept up to date with the proper software. This helps you to focus more on your business' core aspects and issues and not let smaller issues sidetrack you from your original goals and business plans.

Let the Employees Benefit, Too

When you are running a business, you are not the only one who is involved. You have to listen to your employees, as well. If a business is a machine, then your employees are the well-oiled cogs that keep it running.

But unlike cogs, their opinions do matter and you cannot successfully utilize the Cloud without getting your employees on board with the chosen program. However, you do not have to worry about a revolt happening over the utilization of the Cloud. Therefore, it is best if you take the time to get their opinions, too. In the end, they are the ones who will be interacting with the Cloud on a more personal and regular basis. If they are unable to conform to the Cloud, then you might have some difficulties on your hands.

Yet, you should not fear at all. Cloud Nine is a loveable thing. It provides so many benefits for the end users that it is impossible not to love and enjoy working with it. Regardless of what kind of device is used to access the database within the Cloud, the results that Cloud based apps produce are the same.

With an easy to access and easy to understand user interface, the Cloud will become such an ingrained part of your employees' work routine that you will wonder how you ever got anything done without it.

BYOD!

This increase in productivity can be kept up if you start introducing the BYOD (Bring Your Own Device) work ethic which is becoming popular among corporate offices.

The BYOD program has been embraced with great enthusiasm since it allows employees to bring the device they are most comfortable with to work. These could be anything from laptops to tablets to smart phones.

In the end, the BYOD program increases productivity while lowering the overall costs of the business. You will no longer have to spend so much money trying to buy computers in bulk. Just let them bring their own devices.

Where does the Cloud fit in with such a program? Well, the Cloud allows the employees to access your database by going through a login screen.

 In this way, there is increased security and you will be aware of who is logging in and who is not. With the help of such programs and the Cloud, you will be able to make your cloud grow big enough to make it rain.

Look To the Horizon!

Now that you have the cloud easily incorporated with your business, it is time to look at the kinds of things you can now explore with the help of your Cloud. Do you want to save on costs and make your business smoother? You can finally do that with the help of the Cloud. Up-front costs can drastically decrease and you can now do that with ease.

Incorporating the Cloud into your business can help you to take advantage of having a smaller work force. Even if you want to store files, you can do so easily without wasting energy and precious resources on it.

Once you have selected the Cloud of your choice, feel free to take out your list of items regarding the expansion of your business and start drawing up plans for what to do next.

Do you want to start a virtual office for your employees where they can just log in from anywhere and work on their documents? Working on these same lines, Cloud computing also helps you to create a collaborative environment within the office. This means that your employees will be able to work on joint projects with more efficiency. Furthermore, if you introduce a project that requires the help of freelancers, they can just log into the Cloud with the help of your secure log in credentials.

This will minimize the requirement of having to email your files and folders back and forth every time some changes have been made to the documents. Be prepared for the increase in business, productivity and profits. The sky is the limit with the help of the proper Cloud and, once you have your Cloud Nine, you can grow big in ways that can literally make it rain.

Chapter 10 - Landing in Seventh Heaven with Cloud Nine

Dearest Reader,

We have finally come to the conclusion of our epic journey and it was a pleasure showing you everything there is to know about the Cloud.

We hope we have accomplished the goal of providing you with the information you require in order to make the best and most educated choice when it comes to utilizing a Cloud for your business.

If you follow the tips and use a little ingenuity, you will be able to incorporate the Cloud in such a manner as to make it a viable addition to your business practices. Even if you are just curious about getting to know the Cloud more, we hope this book will definitively help provide you with the information you need or want about the Cloud.

Once your realize all the potential, the ease of utilization and the whole world of possibilities that your Cloud opens up for you, it will be safe to say that you will be on the path to reaching Seventh Heaven with the help of your trusty Cloud Nine.

However, the Cloud is something which is constantly growing and evolving according to changes in technology. It is best to keep an ear to the ground and pay attention to all the changes that are taking place within the economy. While businesses have been slow to adapt, the changes are happening much more rapidly than ever thought possible.

Remember, those who are left behind are often forgotten but if you stay up to date and keep in touch with the market place, you will be able to change your sails to catch the winds of change before they sweep you over and make you capsize.

In the end, it is impossible to say when you will reach Seventh Heaven. Human nature is never satisfied with what they have and it is normal to reach forward for more. Furthermore, with the Cloud still showing possibilities of developing and evolving further, we will leave it up to you to recognize and realize when you have reached Seventh Heaven.

TRY NOT.
DO
OR DO NOT.
THERE IS NO TRY.
– YODA

Goodbye and good luck!

An Invitation to the Reader

The reason we published this book was to fortify business owners with the basic knowledge they need to make a great decision when choosing a cloud solutions. We believe a qualified computer consultant can contribute to your business success just like a great marketing consultant, attorney, accountant or financial advisor.

The technology industry is so new, and growing at such a rapid pace, that most business owners cannot keep up with all the latest whiz-bang gadgets, alphabet soup acronyms, and choices available to them. Plus, many of the "latest and greatest" technological developments have a shelf life of six months before they become obsolete or completely out-of-date. Sorting through this rapidly moving mess of information to formulate an intelligent plan for growing a business requires a professional who not only understands technology and how it works, but also understands how people and businesses need to work with technology for progress.

Unfortunately, the complexity of technology makes it easy for a business owner to fall victim to an incompetent or dishonest computer consultant. When this happens, it creates feelings of mistrust toward all technology consultants and vendors, which makes it difficult for those of us striving to deliver exceptional value and service to our clients.

Therefore, our purpose is to not only give you the information you need to find an honest, competent computer consultant, but in doing so, to raise the standards and quality of services for all consultants in our industry. We believe that the more this topic is discussed, the better it will become for all involved.

We certainly want your feedback on the ideas in this book. If you try the strategies we have outlined and they work, please send me your story. If you have had a bad experience with a computer consultant, we want to hear those horror stories as well. If you have additional tips and insights that we have not considered, please share them with us. We might even use them in a future book!

Again, the more aware you are of what it takes to find and hire great consultants in every aspect of your business—not just technology—the stronger your business will become. We are truly passionate about building an organization that delivers uncommon service to our customers. We want to help business owners see the true competitive advantages technology can deliver to your business and not just view it as an expensive necessity and source of problems.

Your contributions, thoughts and stories pertaining to my goal will make it possible. Please write, call or e-mail us with your ideas. info@tsiva.com

Book Order

If you enjoyed this book, share it with others! Use this form to order extra copies for friends, colleagues, clients, or members of your association. Please allow two to four weeks for delivery.

Quantity Discounts:
1-9 copies = $19.95 (Discount from $24.99) each
10-49 copes = $16.95 each
50-99 copies = $13.95 each
100 or more copies = Call for discounts and wholesale prices

Information:
Name:_____
Company:_____
Address:_____
City:_____
State/
Province:_____
ZIP/postal code:_____
# of copies_____@$_____Total: $_____

Add shipping and handling @ $3 per book: $_____
Please make check or money order payable to:
TSI – Your Technology Partner
Credit Card: ☐ Visa ☐ MC ☐ Amex ☐ Discover

Number:_____
Expiration Date:_____
Security Code (3 Digit on back of card):_____
Signature:_____

FAX to: 703-637-1284
Email to: sales@tsiva.com

TSI – Your Technology Partner
3320 Noble Pond Way Suite 201
Woodbridge, Virginia 22193 Phone: 703-596-0022

Made in the USA
Lexington, KY
30 July 2017